YOUR KNOWLEDGE HAS VALUE

AF167107

Plant Design and Manufacturing Principles in DNA Vaccine Production

Sharyar Ahmed

Bibliographic information published by the German National Library:

The German National Library lists this publication in the National Bibliography; detailed bibliographic data are available on the Internet at http://dnb.dnb.de.

ISBN: 9783346248251
This book is also available as an ebook.

© GRIN Publishing GmbH
Nymphenburger Straße 86
80636 München

Print and binding: Books on Demand GmbH, Norderstedt, Germany
Printed on acid-free paper from responsible sources.

The present work has been carefully prepared. Nevertheless, authors and publishers do not incur liability for the correctness of information, notes, links and advice as well as any printing errors.

GRIN web shop: https://www.grin.com/document/903166

PLANT DESIGN AND MANUFACTURING PRINCIPLES IN DNA VACCINE PRODUCTION

By Sharyar Ahmed

MARCH 9, 2017
UNIVERSITY OF BIRMINGHAM

Abstract

The demand for DNA vaccines in large quantities at high purity for gene therapy is on the increase. As it helps to stimulate antibodies production in human and provide immune protection against many diseases such as cancer, malaria, HIV and other diseases (Laere et al., 2016) and have potential advantages over conventional vaccines.

Therefore, this report covers a detail design and cost (to an accuracy of +/- 20%) for a new manufacturing facility to produce DNA vaccines to be built on a greenfield site. Applied current good manufacturing practice (cGMP) and complied with all the regulatory guidelines set up by various agencies.

A process to produce commercial pDNA covers: Fermentation, recovery and product purification. The layout of the manufacturing facilities has been designed in a way to allow a good waste, raw material and personnel flow to minimise the risk and contamination. All of these are an example of current good manufacturing practices which are vital for the production of the therapeutic product (DNA vaccines).

Plant services, system and utilities has been designed to meet the requirement for the manufacturing facility which includes Water for CIP to clean equipment and media preparation, Clean steam for SIP to sterilise the equipment, Heating, ventilation and air conditioning (HVAC), water, Compressed air, Effluent treatment, Nitrogen, USP, CIP system, refrigeration and more. (Cole, 1998).

The facility will take up to 2.6 years to complete and qualify (GSK, 2017). The project schedule will depend on various activities these must be completes to produce DNA vaccine facility, ready for start-up. These includes Front end design, Detailed design, Procurement of equipment, construction, recruitment of staff, commissioning of facility and validation of process. (Adkin, 1998)

Finally, the costs of the pharmaceutical plant are a serious consideration when planning whether a plant will be commissioned or not. The economic evaluation for the DNA Vaccine Plant suggests that the plant will be at significant profit with a payback time of less than 2 years.

Contents

Plant design and manufacturing principles in DNA vaccine production

1.0 Introduction

DNA Vaccines helps to stimulate antibodies production in human and provide immune protection against many diseases such as cancer, malaria, HIV and other diseases (Laere et al., 2016) and have potential advantages over conventional vaccines. The demand for DNA vaccines in large quantities at high purity for gene therapy is on the increase. However, inaccessibility of DNA vaccines for the treatment of diseases can cause several problems. Therefore, to meet the rising demand for DNA vaccines, there is a need for designing & manufacturing facilities to produce DNA vaccines. In order to fulfil the need for the production of DNA vaccines the manufacturing facility such as design, operation and layout must comply with the regulatory requirements stipulated that the production of DNA vaccine must be performed under current good manufacturing practices (cGMP) and good design practices (GDP). Also, comply with the guidance regarding the specification, quality testing and manufacturing standard of DNA products by regulatory agencies such as the US Food and Drug Administration (FDA) and the European Agency for the Evaluation of Medicinal Products (EMEA) and also regulation stipulated by the country where the facility is being constructed. (Przybylowski et al., 2017)

According to the EU guide, GMP is the part of Quality Assurance (QA) which make sure that products are consistently produced and controlled to the quality standards appropriate to their intended use and as required by marketing authorisation or product specification (Institution of Chemical Engineers (Great Britain), 2003).

When designing a new drug production manufacturing facility, the following consideration needs to be made (GMP requirements). To ensure the safety, identity, purity and potency of the manufactured product (Przybylowski et al., 2017). Not placing the patient at risk due to insufficient safety or quality.

- Establishment of effective QA system
- Control of the process
- Personnel that are competent, appropriately qualified and trained.
- Facilities – suitable for operation, layout, design and operation reduces the risk of errors. Cleaning in place (CIP)
- Premises and equipment that are critical for product quality are validated (qualified).
- Standard operating procedure to Prevent contamination from any sources such as environment, equipment and premises.
- Environmental impact assessment (EIA)
- Documentation and audit of all aspects of the process.

1.1 Aims/objective

- Propose concept designs and costs (to an accuracy of +/- 20%) for a new manufacturing facility to produce DNA vaccines to be built on a greenfield site. Apply current good manufacturing practice (cGMP) and comply with all the regulatory guidelines set up by various agencies.
- Schedule for the project / timeline of the project.
- The facility design will include – Design of manufacturing facility, manufacturing flow, plant services system and utilities, Process validation & cGMP and costing.

2.0 Detail of the process

A process of commercial pDNA manufacturing includes: Fermentation, recovery and product purification. Below figure 1 shows the unit operation involved in the manufacturing process.

2.1 Process description

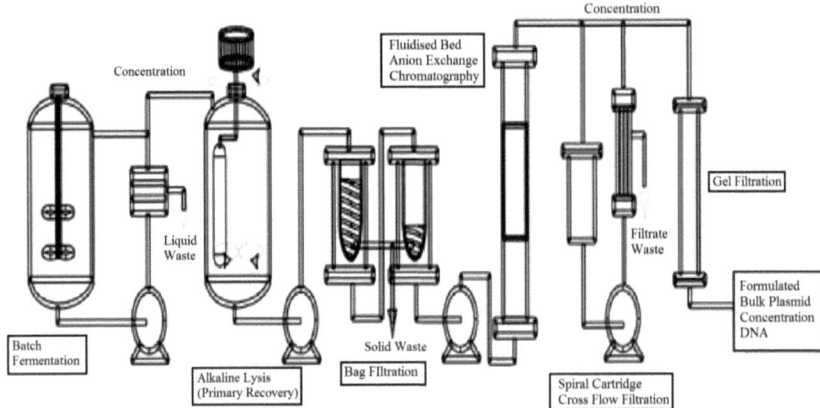

Figure 1 : the process flow of the production of therapeutic plasmid purified bulk products. Adkin, A. (1998). Design of a multi-use DNA vaccine production facility. *Project report submitted for partial fulfilment for the degree of Master of Science in Biochemical Engineering.,* [online] 1(UCL). Available at: https://www.researchgate.net/publication/260581752_Design_of_a_multi-use_DNA_vaccine_production_facility [Accessed 5 Mar. 2017].

<u>Batch Fermentation</u> – Fermentation process is the starting point for the production of pDNA in the bio-reactor, and the aim is to maximise pDNA productivity per unit cell mass. This takes place under the optimum condition to maximise the cell growth. (Adkin, 1998)

Received raw material is placed in quarantine unit until clearance from QA. Production culture stains within the facility are maintained as MCB and WCB's. At the start, bacteria from a WCB's vial is revived and grown in a propagation area up to 2 cultures. Sterilised prepared media is feed into the fermenter. Two seed vessel of volume 15L and 150L are used to produce inoculum for the 1000L cell culture vessel. Operated in the fed-batch mode to provide product for successive downstream processing with operating temperature of 37°C optimum for bacteria. Upon the establishment of maximum cell growth. The broth is cooled to 4°C and pumped to centrifuge at a flowrate of 1000L/h, as this allows effective separation of cellular material. Concentrated broth contains pDNA and some other cell organelles; this is discharged in the alkaline lysis vessel. (Adkin, 1998)

4

Alkaline Lysis – three step process. 1) concentrated broth is suspended in the buffer solution for alkaline lysis - low viscosity. 2) The cell suspension is lysed by mixing with NaOH/SDS – solution changes highly viscous. 3) 3M potassium acetate neutralisation causing the precipitation of impurities such as chromosomal DNA whilst minimising the loss of pDNA yield, the solution becoming Newtonian again. Bag Filtration – Disposable non-leaching nylon bags doesn't contaminate the product, fully scaleable. This partially removes the solid precipitate from the lysate before further product purification.

Anion Exchange chromatography – separation occurs due to electrostatic attraction between the solute and charged dense clusters. Anion exchange chromatography clears all the cationic proteins and lipopolysaccharides and retains all forms of pDNA, the residual cDNA, RNA and any anionic proteins. (Nigel, 2017)

Gel filtration – The removal of small RNA, aggregates or clipped products by Size Exclusion Chromatography; a non-adsorptive chromatography process that separates cluster that differs in size from pDNA in desalting solution. (Nigel, 2017)

Endotoxin Removal – the produced product should be free from all sort of contaminants for the purposes of therapeutic. Therefore, the final step includes the removal of anionic endotoxin by a particular adsorptive chromatographic process. (Nigel, 2017)

Finally, formulating the product by putting stabiliser, correct salt concentration and sterile filtration & lyophilisation.

2.2 Raw material required

The following Raw materials are needed for the production of DNA vaccines. All the raw materials entering the process should be controlled and tested to make sure the manufacturing process meet the standard requirements. The most important Raw material is plasmid DNA. Therefore, it should be fully characterised. For Therapeutic pDNA E. coli is the most common host. (Przybylowski et al., 2017)

Raw materials required – pING human tyrosinase (pING-HT) plasmids, E.coli, Nutrient - Glucose, antifoam,water, liquid nitrogen, ammonia hydroxide, Ethylenediaminetetra acetic acid diammonium salt (EDTA), Hydrochloric Acid, Magnesium Sulphate, Polypropylene Glycol (PPG), Potassium Acetate, Potassium Phosphate, Sodium chloride, Sodium Hydroxide and Tris Hydroxymethyl methylamine (TRIS) buffer. (Adkin, 1998)

2.3 Equipment required

The following table shows the manufacturing equipment required for the process, these should be designed, located and maintained to suit its intended purpose and these should not affect the quality of the product.

Number	Equipment	Size
2	1st Seed Fermenter with motor	15L
2	2nd Seed Fermenter with motor	150L
2	Production Vessel with motor	1000L
9	Antifoam, Base and Acid reservoir	10L
3	Air compressor	20/200L/min
1	Boiler Unit	250kW
1	Chilling Unit	400kW
1	Bag Filtration	50L
1	Disc stack centrifuge	1000L/h
1	Expanded Bed Anion Exchange column	42L
1	Tank (alkali lysis)	4000L
3	Lysis reagent mixing tanks	1500L
1	Ultrafiltration unit	$9m^3$
1	Gel Filtration column	150L

Number	Equipment	Size
1	Gel Filtration purified fraction tank	50L
1	Ultrafiltration purified fraction tank	50L
2	Anion Exchange buffer tanks	300L
1	Anion Exchange purified fraction tank	300L
1	Sodium Chloride Tank	300L
1	Ultrafiltration disposal fraction tank	600L
1	Gel Filtration disposal fraction tank	600L
1	Lysate Holding tank	4000L
2	Kill tank	$6m^3$
15	Pumps	L/hr

Table1: Equipment used in the process (Adkin, 1998)

2.4 Staff Involved

The correct manufacturing of the product (DNA Vaccines) depends upon people/staff. Therefore, qualified personnel must carry out all the task which are the responsibility of the manufacturer. The manufacturer should have enough staff with essential qualification and experience, and they should be aware of cGMP and receive appropriate training. (Inspection and Healthcare, 2002)

All the staff members whose activate could affect the quality of the product should be provided with adequate and continuing training such as technical, maintenance and cleaning staff.

Key personnel include Head of Production and Head of Quality Control (QC).
Other personnel include Process Engineer, Technician, Maintenance, Operative, Assistance and cleaning. (Inspection and Healthcare, 2002)

3.0 Design of Manufacturing facility

The design of the new manufacturing site for (DNA Vaccine) will be laid out for primary production, secondary production, research and development, warehousing and distribution/administration and head office activities. (ICheme, 2003)

3.1 Plant location

The location of the drug production manufacturing facility is crucial to its profitability as it can be affected by raw material supply, climate, transportation, utilities and services, etc.

Site selection

The DNA vaccine plant will be located in southern England, Slough. This is chosen because the site is relatively near to greater London which can create huge advantages which include:

- Transport – the site is located by M25, which will allow cost effective transportation. The site is also located near to the Heathrow airport; this will allow cost-effective exports (Products) and imports (Raw materials) while timesaving.
- Labour - The site is relatively near to greater London which has high-density population. Therefore, there will be plentiful labours for the plant, and there is good transport link.
- Site cost – as the land is outside the vicinity of London, the plant cost is very low.
- Utilities and services – the plant site is near to lots of different water supplies such as a river (Thames) and lakes. As the facility, will have high power and steam requirement, this will be available from various sources such as generators and turbines.

3.2 Plant layout

The detailed design of the layout of the Plant site and Manufacturing facility are given in the appendix. The layout of the manufacturing facilities will be designed in a way to allow a good waste, raw material and personnel flow to minimise the risk and contamination. All of these are an example of current good manufacturing practices which are vital for the production of the therapeutic product (DNA vaccines). The design of manufacturing facility is divided into three sections which include:

- Process Areas – part of facility involve is production process. Includes Fermentation room, alkaline lysis room, purification room, filtration room, raw material room, cell bank, CIP/SIP room and preparation room. (GSK, 2017)
- Process support areas – additional requirements that are important for process operation while following cGMP's. includes inspection area, a decontamination area, Mobile access lobby, personnel access lobby, ancillary equipment room, packaging and labelling area, conveyor facility. (GSK, 2017)
- Amenities – areas on the site and on the manufacturing facility includes changing rooms, car park, reception desk, security, canteen, etc. (GSK, 2017)

3.3 Manufacturing Flow

Material/Waste flow

This includes all the movement of materials or waste material. Typical criteria include: (ICheme, 2003)

- Materials flow through the area, for example, linear flow through with no cross-over of production streams.
- Methods of handling and prevention of cross-contamination.
- Storage condition (refrigerated, toxic, hazardous, filtered) (ICheme, 2003)

Raw Materials and supplies flow

Quarantine storage room --- Release Storage room (wipe with 70% isopropanol) --- Gowning room (wipe with 70% isopropanol) --- pDNA production Cleanroom.

Final Product flow

 pDNA production cleanrooms --- Gowning room ---- Final-product storage room

Waste flow

 pDNA production cleanrooms ---- Gowning room --- Biomedical waste disposal

Personnel Flows

This includes the influence personnel have on the quality of the product that might be caused by their contact with the product. The criteria that have been included in the Production House layout include: (ICheme, 2003)

- Clothing Requirements – (Changing rooms and gowning at the beginning and at each production area).
- Security and access control including potential short cuts and back doors.
- Potential points of cross-contamination between personnel such as changing rooms and gowning room. (ICheme, 2003)

Personnel flow example

Pre-Gowning room to Gowning room (Sterile garments, double gloves) to pDNA production cleanroom.

The layout of the Production house illustrates the Personnel flow, Material flow, waste and equipment flow appendix 2. Different pathways for clean and unclean materials/Personnel to avoid mixing and contamination has been used in the design to comply with cGMP requirements.

4.0 Plant Services, Systems and Utilities

Plant services, system and utilities are designed to meet the requirement for the manufacturing facility. DNA Vaccine manufacturing requires following services:

Water for CIP to clean equipment and media preparation, Clean steam for SIP to sterilise the equipment, Heating, ventilation and air conditioning (HVAC), water, Compressed air, Effluent treatment, Nitrogen, USP, CIP system, refrigeration and more. (Cole, 1998)

4.1 Heating, ventilation and air-conditioning (HVAC) system

HVAC system is an important part of the production clean rooms design and layout, ensuring that the product purity and quality are not affected by the room temperature, humidity, pressure, air particles and cross contamination. Providing control/healthy environment for process operation and occupants. (UK, 2015)

HVAC system is managed by automated temperature control system (ATC), which controls the humidity, Air distribution and air quality for the process operation. There various types of HVAC systems which are found in the pharmaceutical facilities and selection depends on the requirements within the plant such as environmental conditions and the level of product containment. The degree of control within the plant increases the complexity and the cost of HVAC. (ICheme, 2003)

The HVAC systems for the DNA vaccine manufacturing facility will include: Separate systems for each work centre and total loss systems to minimise the risk of cross-contamination. Terminal HEPA filters on supply and extract to control dust, microorganisms and airborne particles. Sterile (+) or containment (-) pressure cascades by Airlocks and air handling unit (AHU), Low humidity, Dust extract and specified classification of the clean room. (ICheme, 2003)

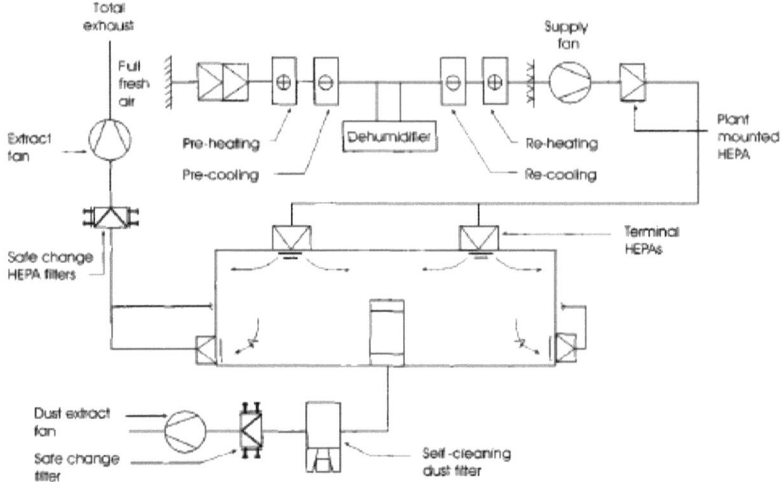

Figure 3: Low humidity total loss containment clean room. Cole, G., Bennett, B. and ICheme, (2003). Engineers guide to pharmaceuticals production - IChemE. Rugby: The Institution of Chemical Engineers.

4.2 Water

The DNA vaccine industry comes under the regulations of MCA, FDA and MHRA. They constantly review with great attention the treatment of water in all segments of the industry used for rinsing, washing and product formulation. (Cole, 1998)

Water is considered to be one of the most difficult product to maintain the required standard not only because of the chemical impurities but, the bacterial contamination, as some system contain components which support the growth of bacteria in water (Cole, 1998). DNA vaccine manufacturing facility will use the water provided by the water treatment plant. Town water, process water and wastewater from the plant will be treated to provide other types of water by various processes such as ion Exchange Deionisation (DI), Reverse Osmosis (RO), Electro Deionisation (EDI) and Distillation (WFI). (Cole, 1998) Types of water will be used in the manufacturing facility are:

<u>Town water/portable water</u> – water from the mains and the quality changes throughout the year. Also, known as the drinking water. Microbiological less than 500cfc/ml and no bacterial colonies according to WHO. (ICheme, 2003)

<u>Process Water</u> – this is normally a portable water which has been passed through the site. Used for washing equipment and cooling. This is also an input for the purified stream (ICheme, 2003)

Purified Water – softened and passed through a UV source to kill live bacteria. Preparation of drugs that do not require WFI (non-sterile products), development of clean solution and input for WFI. According to WHO, Microbiological level less than 100cfc/ml and without any pathogens. (Cole, 1998)

Water for injections (WFI) – softened, extremely low bacterial count and reduced endotoxin. Commonly used for the preparation of the sterile drug. This is generated via distillation with some other treatment method (Cole, 1998). Aerobic Bacteria level < 10cfc/ml and Endotoxin < 0.25 I.U./ml.

The town water and the process water will be used in amenities – admin, lobby, drinking and washing equipment and cooling. The purified water and WFI will be utilised in the main production line for producing the drugs. Purified for initial rinsing and WFI for final rinsing. (GSK, 2017)

Purified water process flow for DNA Vaccine Facility
The following process will be used to purify the town/process water for the production of DNA Vaccine.

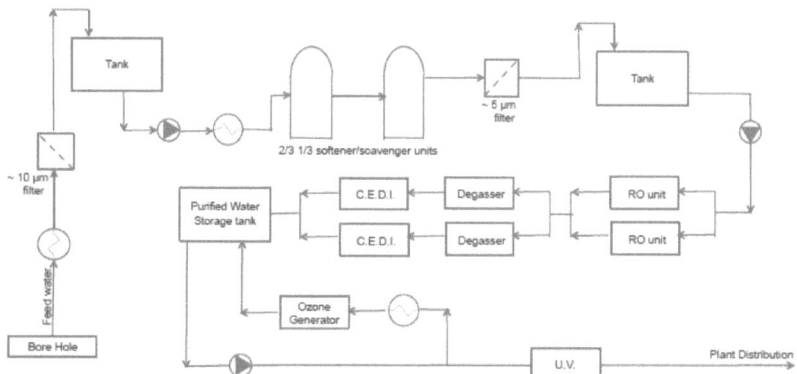

Figure 4: Purified water process flow. Vilhelmsen, B. (2017). Water Systems Challenges for the Pharmaceutical Industry. MDPI Technologist: GSK.

4.3 Clean steam
Clean steam is produced using a steam generator, and it is commonly utilised in a pharmaceutical application where the steam and its condensate are in direct contact with the product to avoid contamination (this is governed by cGMP). (ICheme, 2003)

Clean steam will mainly be used for the sterilisation of process as a sterilising agent; it will also be used in an autoclave and for the humidification for clean rooms. The steam must have the same specification as the water for injections by BP standards for WIF, ensuring the right quality and cleanliness of the steam. This requirement allows the WIF for steam generation. (ICheme, 2003)

Pipeline within the facility can trap condensate which poses the risk of bacterial growth and reduces the effectiveness of sterilisation. Therefore, there should be a steam trap at 30m intervals for efficient removal of condensate. (ICheme, 2003)

4.4 Heat and power

Combined heat & power CHP station will be used onsite to provide heat and power supply. CHP is also known as cogeneration simultaneous production of electrical power and heat from natural gas, and it has a heat recovery system, uses the by-product heat which results in higher thermal efficiency over the conventional power station. The overall efficiency of CHP system is greater than 80%. (admin, 2014)

Therefore, the facility will use CHP station onsite to provide heat and power supply. CHP will operate on natural gas, biomass and waste (as a renewable source). The facility will use the excess thermal energy for steam generation and for heating by using pinch technology.

4.5 Cleaning Systems

It is critical that the DNA vaccine manufacturing facility to obey with all the GMP requirements. The level of hygiene is one of the primary element, involves cleaning plant items and areas that can affect the purity/quality of the manufactured product. Clean in Place (CIP) and sterilise in Place (SIP) will be used in the facility to minimise product loss, removal of soil from product contact surfaces and elimination of micro-organism to meet the desired hygiene standard. But, validation of the cleaning procedure will be necessary before it can be applied. Once this is approved by the relevant governing body, the maintenance of the cleaning systems should be continuously monitored throughout the cleaning and documented which includes the parameters such as flowrate, time, temperature and detergent.

4.51 Clean in Plane (CIP)

CIP is a very simple system, consist of a tank filled with the correct concentration of cleaning agent which is heated to a specified temperature by recirculation and then this is pumped into the pipework or vessel followed by drainage. The important consideration here is the control of flow rate, superficial velocity, temperature and the concentration of detergent of cleaning agent. The facility will use decentralised maximum detergent recovery system. The integration of CIP system with process plant to allow unit cleaning as it has great flexibility, efficiency and objects can be cleaned during production. The cleaning will be measured in different ways such as measurement of mass/volume to assure the right amount of CIP fluid gone through; Pressure ensures that fluid is contacting with the surface at the right force and Conductivity to make sure that the chemical is out of the system. (ICheme, 2003) (Goode, 2017)

Function of typical CIP stages

Stage	Major components added	Function
Prerinse/ Product recovery	Water - usually warm	Removal/recovery of residual product and loose soil
	Solid 'pig'	Product recovery using tight fitting rubber/ plastic pig forced through line to remove residual product
Detergent	Alkali - usually hot	Solubilisation and removal of organic components
	Acid - usually cold	Removal of inorganic soil
Intermediate rinse	Water - hot/cold	Remove detergent residues from line
Disinfection	Water - hot Chemical - warm/cold	Reduce level of micro organisms to level compatible with desired hygiene standards
Sterilisation	Hot water - pressurised Steam - pressurised Chemical	Destruction of all micro organisms
Final rinse	Water - cold	Removal of detergent/disinfectant residues [26]

Table 2: Typical function of CIP stages (Goode, 2017)

4.52 Sterilise in Place (SIP)

SIP is not a part of CIP, and this usually comes after the CIP system as detergent cannot clean all part of the plant. The aim of the SIP system is to kill all the viable microorganism (sterilise). Thermal sterilisation (saturated steam sterile) will be used for sterilising the plant equipment which has already been cleaned using CIP system. Similar paraments will be monitored for SIP measurements (Goode, 2017). Steam for cleaning must be pure, and the pressure must be down to 1.2 Barg and temperature 121°C at the point of use.

4.6 Cleanrooms

To produce a sterile product (DNA Vaccine), clean environment is required to minimise the risks of product contamination. In the manufacturing process, people and manufacturing personnel are the biggest sources of contamination, therefore, to reduce the contamination a clean room need to be designed. A clean room is a room in which air supply, air distribution and air filtration are regulated to controls the level of airborne particle concentrates and also meet the cleanliness standards specified by the current standard ISO 14644-1. There is also a specific cleanroom classification for pharmaceutical manufacture figure 4. The table illustrates that the clean rooms are classified into various grades, and it is specified by EU GMP Annex 1. (ICheme, 2003)

GRADE	AT REST		IN OPERATION	
	Max. number of particles/m³			
	≥0.5 µm	≥ 5 µm	≥ 0.5 µm	≥ 5 µm
A	3,500	0	3,500	0
B	3,500	0	350,000	2,000
C	350,000	2,000	3,500,000	20,000
D	3,500,000	20,000	NA	NA

Figure 5: cleanroom categorisation based on max no of particles/m³

There are 4 grades of the clean room.

Grade A – the local zone for high-risk operation, Grade B: for aseptic preparation and filling and grade C & D: for clean areas for carrying out less critical stages in the manufacture of the product. The cleanroom should be monitored regularly during operation in order to control the particulates. (Inspection and Healthcare, 2002).

The manufacturing facility will have clean room from cGMP grade B to grade D as specified in the plant layout figure 2 in appendices. There will be strict environmental measures for all clean room within the facility to control the air supply/distribution, temperature and pressure. (ICheme, 2003)

Cleanroom with cGMP grade C required lots of air at controlled temperature and humidity, the supplied air in the cleanroom will be filtered through highly efficient HEPA filter which can filter 0.3um airborne particulates with the collection efficiency of 99.97%. Cleanroom with grade C, a unidirectional laminar flow air distribution pattern is essential to a vertical air flow of moderate velocity, where the air is fully filtered via HEPA filters at the ceiling. This type of flow is very useful as it is easier to predict particle travel and no dead areas for contamination. AHU system will be also used to recirculate 80% of the air in the cleanroom to reduce particulate contamination with the product. Also, HEPA filtration on exhaust air system located near the point of extract from the room to minimise contamination. (ICheme, 2003)

Air distribution within in the cGMP grade D will be non-directional turbulent flow; this forms air vortex which affects the removal of airborne particles. (ICheme, 2003)

Isolator technology will be used in the facility to reduce human interventions in processing areas; this will reduce the risk of microbial contamination of DNA vaccine from the environment. And, a minimum number of personnel will be present in the cleanroom, and all employee will receive regular training as this is vital during aseptic processing. (Inspection and Healthcare, 2002)

4.7 Process control & instrumentation

During the manufacturing process of DNA vaccine, certain factors can influence the purity, quality and quantity of the final product. Therefore, effective process and instrumentation control system will be required to maintain the manufacturing process at desired operating conditions, safely and efficiently while satisfying environment and product quality requirement. (GSK, 2017) (AHMED, 2016)

Process and instrumentation control systems monitor and control the process dynamics to function within the set point, set by the operator and the aim of the control loop is to ensure smooth, and rapid return to stable condition following process deviation. Control system also play an important role in plant start-up and shutdown includes alarms and automatic shutdown system this will be integrated part of the system to keep the process variable within safe operating limits. (GSK, 2017) (Ingram, 2017)

The manufacturing facility will use Feedback control, Feedforward control, cascade and Adaptive control as the main control strategy and these will be managed at control room which will be located within the facility. (GSK, 2017)

Feed-forward control will be used in conjunction with the feedback control to measure and correct the effect of process disturbance before they affect the controlled variable. Such as supply of heat to the reactor by a heating jacket. (Ingram, 2017)

Cascade control will also be used. The advantage of having the cascade control is that the secondary measured variable is very close to the potential disturbance. Therefore, cascade control increases the response times, and the potential disturbances are detected rapidly. (GSK, 2017)

Special considerations for DNA products are required such as the flexible operation of process equipment, Batch sequencing /scheduling and avoid cross contaminations. (Ingram, 2017)

Instrumentation

Variable	Instrumentations
Temperature	Thermocouple/RTD
Pressure	Pressure sensor
Air flow	Flow meter
Oxygen or dissolve gas	Membrane sensor
Level	Level sensor
pH	pH probe
Viscosity	Viscometer

Table 3: Controlling variable and measuring instruments (Ingram, 2017)

The purpose of instrumentation is to monitor key process variable during plant operation in order to facilitate process control, process monitoring & warning and process logging. Instrumentation will be designed to monitor relevant process conditions. Onsite (indicators) or in control room (indicate and control). (GSK, 2017) (Ingram, 2017)

In the facility, certain process requires close monitoring, i.e., the aeration of the fermenter (supply of a correct amount of oxygen is essential for the survival of bacteria) therefore, inline and online instrumentation will be used to control the amount of oxygen as this can affect the final output. (GSK, 2017)

There will be three main type of instrumentation in the facility. In-line (placed in direct contact with the product, level/temperature sensor), on-Line (instrument close to process UV analysis) and off-line (sample is withdrawn from the process for analysis, Filter trains). (Ingram, 2017)

5.0 Process Validation and cGMP

The aim of this manufacturing facility is to produce a safe, pure and efficacious product (DNA vaccine), to meet this objective QA procedure must be in place. cGMP is a part of QA which make sure that products are consistently produced and controlled to quality standards appropriate to their intended use. (ICheme, 2003)

When designing the manufacturing facility, the following consideration needs to be made (GMP requirements). To ensure the safety, identity, purity and potency of the manufactured product (Przybylowski et al., 2017). Not placing the patient at risk due to insufficient safety or quality.

- Establishment of effective QA system
- Control of the process
- Personnel that are competent, appropriately qualified and trained.
- Facilities – suitable for operation, layout, design and operation reduces the risk of errors. Cleaning in place (CIP)
- Premises and equipment that are critical for product quality are validated (qualified).
- Standard operating procedure to Prevent contamination from any sources such as environment, equipment and premises.
- Environmental impact assessment (EIA)
- Documentation and audit of all aspects of the process. (Cameron, 2017)

5.1 Process Validation

Guideline on general principles of process validation defines validation as *'Establishing documented evidence which provides a high degree of assurance that a specific process will consistently produce a product meeting its predetermined specifications and quality attributes.'* (ICheme, 2003)

Validation in the Pharmaceutical industry is an important part of a current good manufacturing practice cGMP; it makes sure that the production facility (DNA vaccine Production facility) is performing adequately to meet its specification and quality attributes. Validation is conducted for three main reasons. (ICheme, 2003)

1) Government Regulations - validation protocols of the manufacturing process which includes product specification, selected equipment and processes, criteria for the tests, etc.
2) Assurance of quality – concept of validation is necessary for quality assurance and leads to quality improvement.
3) Cost reduction – reduces waste, rework and rejects, therefore the cost. (ICheme, 2003)

Validation Plan

All the key elements of validation programme such as process evaluation/identification will be clearly defined and documented and will be presented in the form of validation master plan VMP – which defines the overall project validation approach and the associated rationale. Upon completion of VMP, this document will become a valuable tool to present to the regulatory officials that compliance with regulation has been made. The report will contain detail plan describing procedures that have been placed to assure a validated and compliant manufacturing facility with cGMP. (ICheme, 2003). Typical VMP structure is shown below this differs from project to project.

(1) approval page;
(2) introduction;
(3) the aim;
(4) descriptions of:
- facility;
- services/utilities;
- equipment;
- products;
- computer systems;
(5) validation approach:
- overall;
- detail (matrix of validation documents);
(6) other documentation.

Figure 6: VMP structure that will be used for DNA manufacturing facility. (ICheme, 2003)

Requirement & Verification of the Design

User requirement specification (URS) – it's a formal document prepared by the user, record what the user need the design to do, includes specific information on quality, quantity, performance, environment, etc. of the manufacturing product (DNA vaccine). it's the basis for qualification/validation, and it represents a contract between user and project team. Design specification that meets the user requirement is verified by a process called design qualification (DQ). It will be tough to assess whether the design for (DNA vaccine) is suitable or not without URS. (Cameron, 2017)

Qualifications

Validation master plan defines which system are to be qualified, and the next stage involves qualification of equipment and ancillary system before starting process validation. Qualification is conducted by following activities. (Inspection and Healthcare, 2002)

- Design Qualification – verifying that design specification meets all the process, product and user requirements and also, complies with GMP requirements. Fully documented. This will be carried out on various equipment which includes fermenter, anion-exchange chromatography, HVAC systems, centrifuge and filtration.
- Installation Qualification – Documented verification that equipment/systems are correctly installed and comply with design specification recommended by the manufacturer. Such as HVAC systems, process equipment and (steam, CIP/SIP).
- Operational Qualification – Documented verification that equipment/systems function correctly in accordance with design spec. testing divided into component testing & system testing performed using water and compressed air.

- Performance Qualification – Documented evaluation that equipment/ancillary systems connected can produce an output reproducibly to a defined specification. This will be performed on systems whose performance can affect the quality of the product such as purification processes, fermenter, production steriliser, CIP, SIP, WIF, HVAC, PW, centrifuge, etc. as quality control of DNA vaccine is the key component of cGMP. (Inspection and Healthcare, 2002) (Cameron, 2017)

For the production of DNA vaccine, Prospective Validation is the preferred approached, as this is commonly performed on API products and this should be completed before commercial distribution.

lifecycle maintenance

Validation review is the key part of the lifecycle. Therefore, an annual review will be performed looking at aspects of change control and quality data.

What to validate in the DNA manufacturing facility?

- Processes – API stages, intermediate steps affecting API quality and secondary manufacturing.
- Equipment, facility and Utilities – items/system that have a direct impact on the quality of the final product such as purification processes, fermenter, production steriliser, CIP, SIP, WIF, HVAC, PW, centrifuge, etc.
- Computer systems – system that affects the quality and handling GMP data, such as Process control and instrumentation.
- Cleaning – insufficient cleaning affecting the quality/safety of the DNA vaccine such as pipes in the facility. (Cameron, 2017)

DNA vaccine must meet the minimum product specification set by the authority before releasing.

6.0 Gannt Chart

The designing of the manufacturing facility to produce DNA vaccine requires cautious planning and management. The facility can take up to 4 years to complete and qualify (GSK, 2017). The project schedule depends on various activities these must be completed to produce DNA vaccine facility ready for start-up. These includes Front end design, Detailed design, Procurement of equipment, construction, recruitment of staff, commissioning of facility and validation of the process. (Adkin, 1998)

To be competitive in the market, the manufacturing Plant will be constructed, commissioned and validated as fast as possible. To achieve this target, we will use fast track/alliance approach in creating the project schedule and carrying out dependant activates concurrently. (Adkin, 1998)

The Gantt chart shows the time scale of each activity and their time period.

- Front End Design – Submit design to HSE and other authorities. Identify any problem at an early stage and rectify. Time saving and cost. (Adkin, 1998)
- Detailed Design – after front End design. Will allow 8 months. At this stage, the plant layout and dimensions are established. (Adkin, 1998)
- Procurement of equipment – 2 months for user requirement documents to be specified, 12 months for delivery of the items. (Adkin, 1998)
- Construction – as off the fast track approach, the construction will start when at least 20% of the equipment has arrived. (Adkin, 1998)
- Recruitment – Recruitment process will be split into three periods to follow different stages of construction. Senior staff will be recruited at the end of front design will include validation officers, QC and process engineers. Process staff will be recruited when the plant equipment is arriving. The administration staff will be recruited at the end. The requirement will be an ongoing process. (Adkin, 1998)
- Pre-commissioning and commissioning – pre-commissioning will begin soon after when the equipment and machinery have been installed. This process includes checking the dimensions of pipes, specifications of materials, etc. Commissioning will occur when pre-commissioning is completed. this will take 4 months and prepares the plant for OQ. (Adkin, 1998)
- Validation – some of the validation will be carried out at a stage of the vendors such as IQ; protocols will be prepared and checked. OQ, after commissioning and finally the validation department, will perform PQ and PRQ. This process will take up to two months. (Adkin, 1998)
- Start-up – The full phase from design to validation will take 941days, 2.6years.

Gannt Chart Appendix 4

7.0 Costing

Capital Cost Estimate

The cost of the item of equipment can be estimated from the purchase price of the item in previous years. This can be calculated using the Cost index knowledge. (Adkin, 1998)

$$\text{Capital cost present} = \text{Cost index Present} * \text{Capital Cost Past} / \text{Cost index Past}$$

Calculating the current price of the Fermenter 1500L.

Reference Fermenter Size and Price = 1500L and £704509

Price = £704509

1998 Cost Index = 390

2014 Cost index = 19400

Current Price of Fermenter 1500L = 606*704509/390 = £1094698

The full list of estimated delivered equipment cost of the major plant item can be found in appendix.

The total estimated equipment cost of major plant item is calculated to be £3045538.46.

Factor Principle for Capital Cost

Total plant equipment cost C = F*DCE

F = Lang Factor for DNA Vaccine Plant account for additional direct and indirect costs 8

DCE = Delivered equipment cost £3045538.46

Total capital cost C = 8 * 3045538.46 = £24364307.68

Plant costs spread over two years

C = £12182153.84

Working Capital

This is assumed to be 4% of the fixed capital cost

Scrap Value

Assumed to be 15% of the fixed capital cost.

Amortisation

Amortisation is a constant annual payment over the lifetime of the project. This is analogous to a mortgage.

$$A = I * \frac{r(1+r)^n}{(1+r)^n - 1}$$

Where: A= Amortisation, £, I = Initial capital cost, £, r = Interest rate n = number of years to repay the loan.

Assuming the plant life is 12years and the capitals is loaned at 5% interest.

$$A = \mathbf{12182153.84} * \frac{0.05(1 + 0.05)^{12}}{(1 + 0.05)^{12} - 1} = £1374456.502$$

Income

The table below show the current selling per dose and no of doses per year of the product.

Vaccine product	Price per dose	Number of doses produced p.a.	Sales (x 10^3)
Aids	£10	6 million	£60000
Cancer	£7.50	6 million	£45000

Therefore, the total income is £105million per annum, assuming this will be constant.

Operating Costs Estimate

Raw Materials cost

Major materials needed to manufacture the DNA vaccine. Obtained from catalogues. List can be found in appendix 5.

Total cost for the raw material per batch = £49851

Total per year for 282 batches = £14m

Energy Costs

To calculate the energy costs, the energy requirement of the manufacturing process needs to be reviewed. The electricity/utilities costs will be ignored. Total energy cost will be £3566.66. **Calculation in appendix.**

Operating Cost

All the operating Cost calculation can be found in appendix and the final answers are presented below. **Calculation in appendix.**

Type	Cost (£)
Variable Costs	
Raw Materials	14000000
Energy Cost	3566.66
Total	14003566
Fixed Costs	
Labour	584743.2
Plant Overheads	1949144
Supervision	779657.6
Laboratory	779657.6
Insurance	121821
Maintenance	3654646
Tax	121821
Total	7991490.4
Total Operating Cost	21995056.4

Cash Flow Diagram – This shows that the payback time is 1.5years. illustration that the project is feasible.

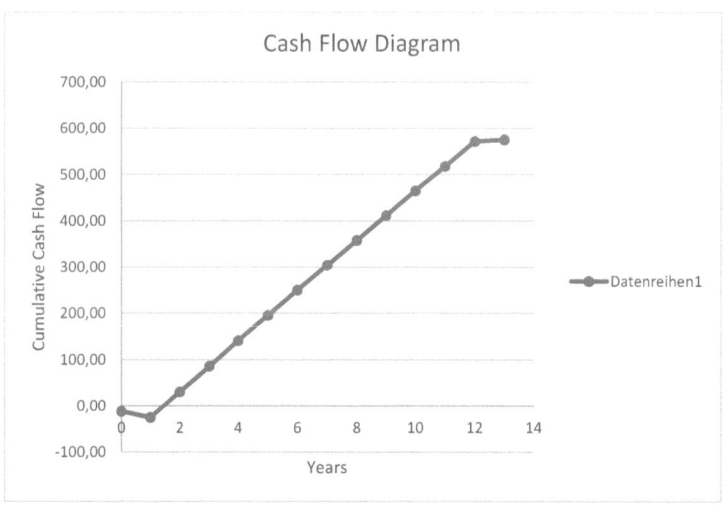

Return on Investment (ROI)

$$ROI = \frac{Cumulative\ Net\ Cash\ Flow}{Initial\ Investment * Plant\ life} * 100$$

This table shows the Return on Investment for the above project.

Net Cash Flow (in millions)	574.50
Plant Life (in years)	12
Capital Cost (in millions)	24.4
Working Capital (in millions)	0.976
ROI as a Percentage	188%

8.0 Conclusion

The demand for DNA vaccines in large quantities at high purity for gene therapy is on the increase. Therefore, to meet the rising demand for DNA vaccines, designed a manufacturing facility to produce DNA vaccines. Applied current good manufacturing practice (cGMP) and complied with all the regulatory guidelines set up by various agencies.

The layout of the manufacturing facilities was designed in a way to allow a good waste, raw material and personnel flow to minimise the risk and contamination as all of these are an example of current good manufacturing practices which are vital for the production of the therapeutic product (DNA vaccines). Plant services, system and utilities was also designed to meet the requirement for the manufacturing facility.

To be competitive in the market, the manufacturing Plant will be constructed, commissioned and validated as fast as possible. To achieve this target, used fast track/alliance approach in creating the project schedule and carrying out dependant activates concurrently.

Finally, the costing of the DNA manufacturing plant shows that the plant is feasible and has significant profit with the payback time of only 1.5years.

9.0 References

Adkin, A. (1998). Design of a multi-use DNA vaccine production facility. *Project report submitted for partial fulfilment for the degree of Master of Science in Biochemical Engineering.*, [online] 1(UCL). Available at: https://www.researchgate.net/publication/260581752_Design_of_a_multi-use_DNA_vaccine_production_facility [Accessed 5 Mar. 2017].

admin, (2014). Combined heat and power to reduce energy costs and emissions. *Chem.* [online] Available at: http://www.engineersjournal.ie/2014/06/05/combined-heat-and-power-to-reduce-energy-costs-and-emissions/ [Accessed 5 Mar. 2017].

AHMED, S. (2016). Advance Process Design.

Cameron, G. (2017). *An Introduction to Good Manufacturing Practice (GMP) & Validation.* 1st ed. Presentation Msc.

Cole, G. (1998). *Pharmaceutical production facilities: Design and applications.* New York: Taylor & Francis.

Cole, G., Bennett, B. and ICheme, (2003). *Engineers guide to pharmaceuticals production - IChemE.* Rugby: The Institution of Chemical Engineers.

Goode, D. (2017). CIP and SIP systems in Cleaning. *Chemical Engineering University of Birmingham.*

GSK, (2017). *Gsk to produce Dna vaccines construction essay.* [online] Available at: https://www.uniassignment.com/essay-samples/construction/gsk-to-produce-dna-vaccines-construction-essay.php [Accessed 5 Mar. 2017].

Healthcare, R. and Inspection, t. (2002). *Rules and guidance for pharmaceutical manufacturers and distributors (the orange guide): 2014.* ed. London: Pharmaceutical Press.

Hussein, S., Azmi, M., Lila, M., Yian, R., Pei, Y., Kiong, A., Ling, and Laere, E. (2016). Journal of botany. *Journal of Botany*, [online] 2016. Available at: https://www.hindawi.com/journals/jb/2016/4928637/ [Accessed 1 Mar. 2017].

Inc, S. (2017). *Pharmaceutical production facilities - design and applications 2nd ed - graham C. Cole (Taylor & Francis, 1998).pdf | pharmaceutical drug.* [online] Available at: https://www.scribd.com/doc/300136141/Pharmaceutical-Production-Facilities-Design-and-Applications-2nd-ed-Graham-C-Cole-Taylor-Francis-1998-pdf [Accessed 5 Mar. 2017].

Ingram, A. (2017). *Process Control in the Pharmaceutical Industry*.

Nigel, (2017). *pDNA manufacture Lecture notes*.

Rivière, I., Sadelain, M., Borquez-Ojeda, O., Bartido, S. and Przybylowski, M. (2017). Production of clinical-grade plasmid DNA for human phase I clinical trials and large animal clinical studies. *Vaccine*, [online] 25(27), pp.5013–5024. Available at: http://www.sciencedirect.com/science/article/pii/S0264410X07004768 [Accessed 1 Mar. 2017].

UK, (2015). *The manufacturing of DNA vaccines engineering essay.* [online] Available at: https://www.ukessays.com/essays/engineering/the-manufacturing-of-dna-vaccines-engineering-essay.php [Accessed 5 Mar. 2017].

Vilhelmsen, B. (2017). *Water Systems Challenges for the Pharmaceutical Industry*. MDPI Technologist: GSK.

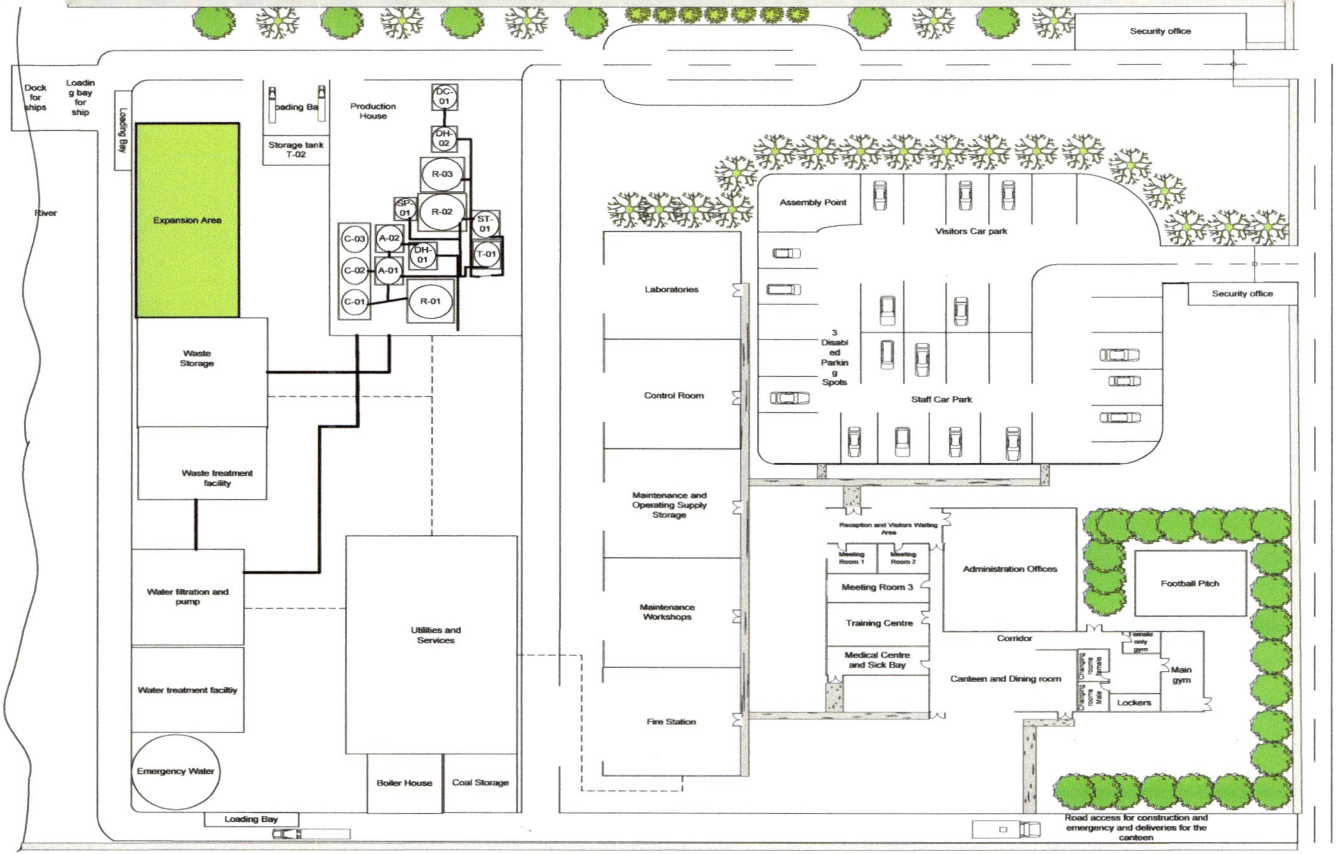
Security office
Dock for ships
Loading bay for ship
Loading Bay
Loading Bay
Production House
Storage tank T-02
DC-01
DH-02
R-03
R-02
SP-01
ST-01
T-01
C-03
A-02
DH-01
C-02
A-01
C-01
R-01
River
Expansion Area
Waste Storage
Waste treatment facility
Water filtration and pump
Water treatment faciltiy
Emergency Water
Loading Bay
Utilities and Services
Boiler House
Coal Storage
Assembly Point
Visitors Car park
3 Disabled Parking Spots
Staff Car Park
Security office
Laboratories
Control Room
Maintenance and Operating Supply Storage
Maintenance Workshops
Fire Station
Reception and Visitors Waiting Area
Meeting Room 1
Meeting Room 2
Meeting Room 3
Training Centre
Medical Centre and Sick Bay
Administration Offices
Corridor
Canteen and Dining room
Changing rooms female
Changing rooms Male
Female only gym
Lockers
Main gym
Football Pitch
Road access for construction and emergency and deliveries for the canteen

Layout of the Production House Appendix 2

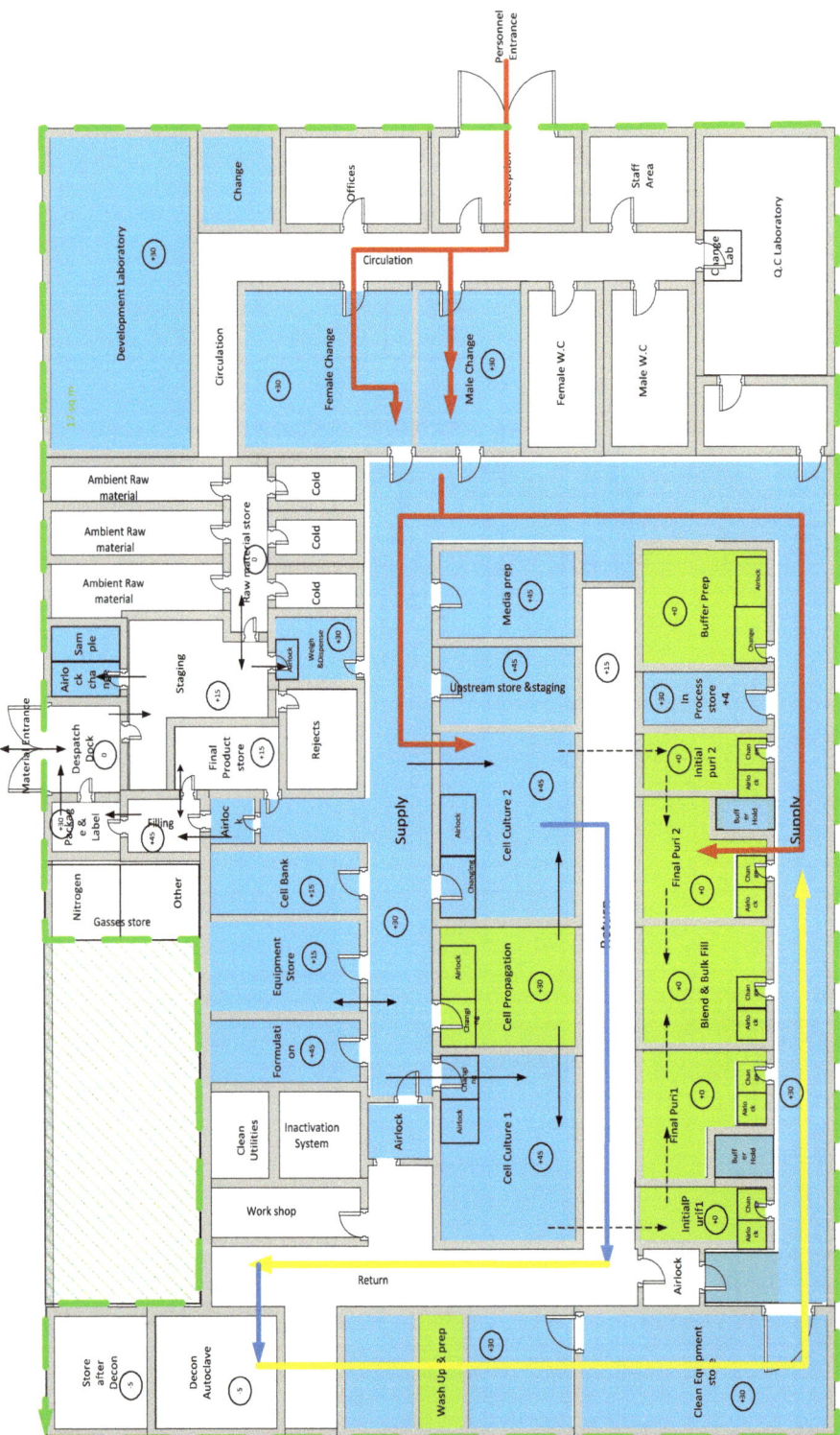

Key

Key for the layout of the Plant site	
Roads	—— ——
Utilities Supply	- - - - - - -
Pipelines	————

Key for the layout of the Plant site	
cGMP grade Unclassified	☐ (white square, green border)
cGMP Grade C	■ (green square)
cGMP Grade D	■ (blue square)
Material Flow (manual Transfer)	→ (black arrow)
Material Flow (piped)	⇢ (dashed black arrow)
People Flow	→ (red arrow)
Equipment and Waste Flow	→ (yellow arrow)
Material and Product Flow	→ (blue arrow)
Facility Area	504m^2 2800 x 1800

ROOM NO: 01

ROOM DATA SHEET

Project Title:	Plant design and manufacturing principles in (bio) pharmaceutical						PDS No:	XXX
Room Name:	PREP ROOM			Proposed Purpose of Room				
Room Dimension:	Width:	3500mm	Length:	6620mm	Height:	2700mm	Area:	23.17m²

Floor	VINYL SHEET LAID ON LATEX SMOOTHING COMPOUND
Finishes	COVED TO WALL (MIN 100mm)
Wall	INTERNAL METAL PARTITION Cleanable
Finishes	VINY SHEET ON PARTITION OR PROPRIETARY CLEAN ROOM PANEL SYSTEM
Ceiling	SEALED METAL (POWDER COATED) SUSPENDED PROPRIETARY SYSYEM
Types	
Door	GRP WITH METAL FRAME
Details	VISION PANEL

Other (Including Fitted Furniture/ Shelves/Display/Equipment Etc.)	
Benching, PC, Autoclave Monitoring System	

Electrical Services

Lighting (Type & Level):	FLUORESCENT 500Lux @ 900mm ABOVE FINISHED FLOOR LEVEL

Power Distribution:	4 X Double Sockets on each wall

Special Requirements:	Protection against water ingress

Cabled Services

Datacom Requirements:	2 X Dataports		
Telecom Requirements:	2 X Telephone ports		
Fire Detection:	Yes	Sprinklers	No
Equipment/ Plant Alarms:	HVAC		
Security Systems:	24 Hour		

Environment

Temperature-Limits:	19 +/- 2C	Humidity- Limits:	Operator
Air Change Rates:	> 20/Hr	Pressure Regimes:	30Pa
Recirc. Air:	TBD	Filtration:	Grade C
Operating Times:	24 Hours	No. of occupants:	Approx.5

Equipment Loads:	Autoclave, Glasswasher
Special Requirements:	Grade C

Local Exhaust Ventilation

Fume Cupboards:	No.				Size:	
Ventilated Cabinets:	No.	1	Type:	Powder Containment Booth	Size:	
Other Vent'd Spaces:	No		Type		Size	

Water Services & Drainage

Sinks:	No:	1	Size:	TBD	Type:	TBD	Taps:	TBD
Wash Basins:	No:	0	Size:	N/A	Type:	N/A	Taps:	N/A
Hose Points:	No:	1	Type:	TBD			Mounting:	TBD
Floor Gullies:	No:	0	Size:	TBD			Type:	TBD
Steam:	Clean Steam, Boiler Steam							
Drainage:	3		Domestic:	No	Sewerage:	No	Effluent:	3

Gases

Compressed Air:	3	CO$_2$	3	Oxygen:	
Nitrogen:		High Vacuum:		Low Vacuum:	
Other:	Continuous Environmental Monitoring Vacuum- non- viable monitoring				

Loose Furniture/ Equipment

Chairs, Trolleys (General), Autoclave trolleys

Equipment to be Installed

Autoclaves, Glasswashers, Column packing station, Buffer Preparation, Laminar Air Flow Units x 2

Approval

Sharyar   ahmed

Prepared by (Signature) _____   Approved by (Signature)_____

SHARYAR AHMED	09/03/17		
Block Capitals	Date	Block Capitals	Date

REVISIONS			
REV	DESCRIPTION	DATE	BY
01	Preliminary Comments Incorporated	9/03/17	SA

Project Title:	Plant design and manufacturing principles in (bio) pharmaceutical					PDS No:		XXX
Room Name:	FILLING ROOM			**Proposed Purpose of Room**				
Room Dimension:	Width:	4090/ 2830mm	Length:	8080mm	Height:	2700mm	Area:	28.616m²

Floor	VINYL SHEET LAID ON LATEX SMOOTHING COMPOUND
Finishes	COVED TO WALL (MIN 100mm)
Wall	INTERNAL METAL PARTITION Cleanable
Finishes	PAINTED BLOCKWORK
Ceiling	SEALED METAL (POWDER COATED) SUSPENDED PROPRIETARY SYSYEM
Types	
Door	GRP WITH METAL FRAME
Details	VISION PANEL

Other (Including Fitted Furniture/ Shelves/Display/Equipment Etc.)	
Work bench	

Electrical Services

Lighting (Type & Level):	SUSPENED FLUORESCENT 500Lux @ 900mm ABOVE FINISHED FLOOR

Power Distribution:	2 X Double Sockets on each wall Surface mounted 3 phase power supply for filling machine

Special Requirements:	Must be classroom compatible

Cabled Services

Datacom Requirements:	Single Dataports		
Telecom Requirements:	1 X ports		
Fire Detection:	Yes	Sprinklers	No
Equipment/ Plant Alarms:	HVAC		
Security Systems:	TBD		

Environment

Temperature-Limits:	19 +/- 2C	Humidity- Limits:	Operator
Air Change Rates:	> 20/Hr	Pressure Regimes:	45Pa
Recirc. Air:	TBD	Filtration:	Grade D
Operating Times:	24 Hours	No. of occupants:	Approx.2
Equipment Loads:	TBD		

| Special Requirements: | | Grade C, suitable for fumigation, LUF unit over autoclave unload. | | | | | | |

Local Exhaust Ventilation

Fume Cupboards:	No.				Size:		Face Vel:	
Ventilated Cabinets:	No.		Type:		Size:		Face Vel:	
Other Vent'd Spaces:	No	1	Type	TBD	Size		Face Vel:	TBD

Water Services & Drainage

Sinks:	No:	1	Size:		Type:		Taps:	
Wash Basins:	No:		Size:		Type:		Taps:	
Hose Points:	No:		Type:				Mounting:	
Floor Gullies:	Yes:		Size:				Type:	
Steam:								
Drainage:	3		Domestic:		Sewerage:		Effluent:	

Gases

Compressed Air:	3	CO$_2$	3	Oxygen:	
Nitrogen:		High Vacuum:		Low Vacuum:	
Other:	Continuous Environmental Monitoring Vacuum- non- viable monitoring 2 off room & within isolator Continuous Environmental Monitoring Vacuum for viable particle monitoring within isolator				

Loose Furniture/ Equipment

Chairs

Equipment to be Installed

Isolator, Filling Machine, Check weight system

Approval

Prepared by (Signature) _____ sharyar _____   Approved by (Signature)_____ ahmed _____

SHARYAR	09/03/17	AHMED	09/03/17
Block Capitals	Date	Block Capitals	Date

REVISIONS			
REV	DESCRIPTION	DATE	BY
01	Preliminary Comments Incorporated	09/03/17	DT

Gantt Chart Appendix 4

ID	Task Mode	Task Name	Duration	Start	Finish	% Complete
1		Front End Design	44 days	Thu 09/03/17	Tue 09/05/17	100%
2		Initial Design	17 days	Thu 09/03/17	Fri 31/03/17	100%
3		Submit to HSE	14 days	Fri 31/03/17	Wed 19/04/17	100%
4		Design Rectify	15 days	Wed 19/04/17	Tue 09/05/17	100%
5		Detail Design	200 days	Tue 09/05/17	Sun 11/02/18	100%
6		Plant Layouts	142 days	Tue 09/05/17	Wed 22/11/17	100%
7		Layout and Dimensions	59 days	Wed 22/11/17	Sun 11/02/18	100%
8		Procurement of equipment	390 days	Thu 11/01/18	Wed 10/07/19	100%
9		User Requirement Document	60 days	Thu 11/01/18	Wed 04/04/18	100%
10		Delivery of items	323 days	Sun 11/03/18	Thu 04/07/19	100%
11		Construction	344 days	Fri 11/05/18	Wed 04/09/19	100%
12		Initial Construction	344 days	Fri 11/05/18	Wed 04/09/19	100%
13		Recurrent	690 days	Tue 09/05/17	Mon 30/12/19	100%
14		Senior Staff	110 days	Tue 09/05/17	Wed 09/10/17	100%
15		Process Staff	94 days	Sun 11/03/18	Wed 18/07/18	100%
16		Administration Staff	172 days	Sat 04/05/19	Mon 30/12/19	100%
17		Pre-commissioning and Commissioning	180 days	Wed 04/09/19	Tue 14/04/20	100%
18		Pre-commissioning	60 days	Wed 04/09/19	Tue 26/11/19	100%
19		Commissioning	100 days	Wed 27/11/19	Tue 14/04/20	100%
20		Validation	380 days	Sun 06/05/19	Thu 15/10/20	99%
21		IQ/OQ	111 days	Sun 05/05/19	Fri 04/10/19	100%
22		OQ/PQ	193 days	Tue 14/04/20	Thu 15/10/20	100%

Project: Simple Project Plan
Date: Thu 09/03/17

Legend:
Task
Split
Milestone
Summary
Project Summary
Inactive Task
Inactive Milestone
Inactive Summary
Manual Task
Duration-only
Manual Summary Rollup
Manual Summary
Start-only
Finish-only
External Tasks
External Milestone
Deadline
Progress
Manual Progress

Costing Appendix 5

The full list of estimated delivered equipment cost

No.	Equipment Item	Size	Equipment Cost £ K
	Fermentation Equipment		
2	SEED VESSELS	15L	68
2	Seed vessels	150L	275
2	Production vessels	1500L	1095
	Down Stream Processing Equipment		
1	BOILER UNIT		171
1	CHILLING UNIT		155
1	DISC STACK CENTRIFUGE	1000L/h	124
1	Alkaline Lysis vessel	4000L	222
1	Expanded Bed Anion Exchange Column	42L	70
1	Gel Filtration Column	250L	48
1	Bag Filter Housing	50L	11
1	Ultrafiltration Unit	9m3	54
1	GEL FILTRATION PURIFIED FRACTION TANK	50L	7
1	Ultrafiltration Purified Fraction Tank	50L	7
2	Anion Exchange Buffer Tanks	300L	21
1	Anion Exchange Purified Fraction Tank	300L	10
1	Sodium Chloride Tank	300L	10
1	Ultrafiltration Disposal Fraction Tank	600L	32
1	Gel Filtration Disposal Fraction Tank	600L	32
3	Lysis Reagent Feed Tanks	1500L	165
1	Lysate Holding Tank	4000L	99
1	Anion Exchange Disposal Tank	5000L	113
2	Kill Tanks	6m3	253
	Total cost of equipment		3045

The total estimated equipment cost of major plant item is calculated to be £3045538.46.

Raw Materials Cost

Major Materials needed to manufacture the DNA vaccine. Obtained from catalogues

Compound	Amount per batch Kg	Cost/unit £/kg	Price per batch £
Ammonia hydroxide	7	728	5096
Antifoam ppg	20	24	480
Glucose	61	0.25	15.25
Magnesium sulphate	1	0.25	0.25
Potassium acetate	1199	0.25	299.75
Sodium chloride 0.1m	1800	0.88	1584
Sodium chloride salt	15	0.25	3.75
Sodium glycine	630	0.25	157.5
Sodium hydroxide 0.2 m + 2% sds	5994	0.88	5275
Tris / edta	6115	6	36690
Tri-sodium citrate 0.1m	1000	0.25	250

Variable Operating Cost

Energy Costs

To calculate the energy costs, the energy requirement of the manufacturing process needs to be reviewed. The electricity/utilities costs will be ignored. Total energy cost will be £3566.66.

The calculations below give an overall energy cost;

Total Energy Requirement: 48.2 MJ

Annual Hours of Operation= 10000 h

Annual Energy Requirement= 428000 MJ

Steam Cost= £0.03/kW

428000 MJ= 118888.89 kW

Therefore,

Energy Cost= 118888.89*0.03

Energy Cost= £3566.66

Fixed Operating Cost

Even in the earliest stages of the design, fixed costs should not be ignored, for they can have a substantial impact on the economy of the project. Fixed costs are the costs that are sustained regardless of the plants operation rate or output. Therefore, if the plant production rate changes, these costs remain the same.

8.3.1. Taxes, Maintenance and Insurance

Taxes and insurance are calculated to be between at 1% and 2% of the fixed capital. Maintenance (material and labour) is estimated to be between 3% and 5% of the investment. Plants that have more solid handling or more moving parts require higher maintenance.

$$Tax = 12182153.84 * 0.01 = £121821$$
$$Maintenance = 12182153.84 * 0.3 = £3654646$$

$$Insurance = 12182153.84 * 0.01 = £121821$$

$$Total = £3898288$$

8.3.2. Labour

Pharmaceutical plants are no longer labour intensive due to high levels of process automation; as such, operating labour costs do not normally exceed 15% of total operating costs.

Total labour costs also include laboratory costs, supervision and plant overheads, which includes security, medical costs, office staff and the canteen.

Laboratory and supervision costs are around 20% of operating labour costs, and plant overheads are around 50% of operating labour costs. Therefore:

$$Assuming\ total\ operating\ costs = 100$$
$$Operating\ labour = 15\% = 15$$
$$Laboratory = 20\%\ of\ operating\ labour = 3$$

$$Supervision = 20\%\ of\ operating\ labour = 3$$

$$Plant\ overheards = 50\%\ of\ operating\ labour = 7.5$$

$$Total\ labour = 15 + 3 + 3 + 7.5 = 28.5 = 28.5\%\ of\ total\ operating\ costs$$

$$100 - 28.5 = 71.5$$

$$operating\ cost = \frac{3898288}{71.5} * 100 = £5452151$$

Cash Flow Table

Plant Life	Fixed Capital costs	Operating costs			Working Capital Cost	Scrap Value	Income	NCF	Cumulative NCF
		Variable (2% due to inflation increase from Year 2)	Fixed	Amortisation					
0	-12.20							-12.20	-12.20
1	-12.20				-0.98			-13.18	-25.38
2		-14.00	-34.00	-1.40			105.00	55.60	30.22
3		-14.28	-34.00	-1.40			105.00	55.32	85.54
4		-14.57	-34.00	-1.40			105.00	55.03	140.58
5		-14.86	-34.00	-1.40			105.00	54.74	195.32
6		-15.15	-34.00	-1.40			105.00	54.45	249.77
7		-15.46	-34.00	-1.40			105.00	54.14	303.91
8		-15.77	-34.00	-1.40			105.00	53.83	357.74
9		-16.08	-34.00	-1.40			105.00	53.52	411.26
10		-16.40	-34.00	-1.40			105.00	53.20	464.46
11		-16.73	-34.00	-1.40			105.00	52.87	517.33
12		-17.07	-34.00	-1.40	0.98		105.00	53.51	570.84
13						3.66		3.66	574.50

All the values are in £ million

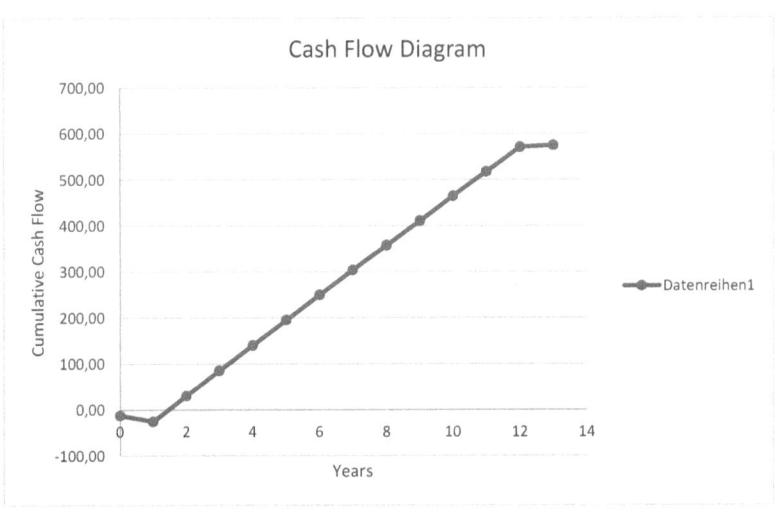